DATE DUE

DEMCO 128-5046

CHARLES DARWIN AND THE THEORY OF EVOLUTION BY NATURAL SELECTION

FRED BORTZ

ROSEN
PUBLISHING®

New York

Published in 2014 by The Rosen Publishing Group, Inc.
29 East 21st Street, New York, NY 10010

Library of Congress Cataloging-in-Publication Data

Bortz, Fred, 1944–
Charles Darwin and the theory of evolution by natural selection/Fred Bortz.
—First edition.
    pages cm. — (Revolutionary discoveries of scientific pioneers)
Includes bibliographical references and index.
ISBN 978-1-4777-1802-5 (library binding)
1. Darwin, Charles, 1809-1882—Juvenile literature. 2. Naturalists—England—Biography—Juvenile literature. 3. Evolution (Biology)—History—Juvenile literature. 4. Natural selection—History—Juvenile literature. I. Title.
QH31.D2B736 2014
576.8'2092—dc23

2013013178

Manufactured in the United States of America

CPSIA Compliance Information: Batch #W14YA: For further information, contact Rosen Publishing, New York, New York, at 1-800-237-9932.

A portion of the material in this book has been derived from *Darwin and the Theory of Evolution* by Robert Greenberger.

CONTENTS

INTRODUCTION

We live on a planet filled with life, with too many species to count. No matter where you look—on land, in the water, in the air, or deep under the ground—and no matter what the climate or environment, some organisms manage to thrive. Where did all those species come from, and how did they manage to find places that suited them so well?

Today we know that species did not go searching for perfect places to live. Rather they adapted so well to the environments where they were already living that they became the ideal creatures for their surroundings. That process of adaptation is what nineteenth-century English naturalist Charles Darwin (1809–1882) called evolution by natural selection.

Darwin first proposed the idea in writing in his 1859 masterpiece *On the Origin of Species by Means of Natural Selection, or the Preservation of Favoured Races in the Struggle for Life*. The book was a result of painstaking research that began when he set sail in 1831 as a twenty-two-year-old naturalist on what was to become a five-year voyage around the world on the H.M.S. *Beagle*.

Nowadays we often write "Darwin's Theory of Evolution" with capital letters to emphasize what the word "theory" really means. As Jerry Coyne writes in *Why*

IN 1859, BRITISH NATURALIST CHARLES DARWIN, SHOWN HERE IN 1878, PUBLISHED A BOOK THAT IS STILL THE FOUNDATION OF THE MODERN SCIENCE OF BIOLOGY. ITS FULL TITLE, *ON THE ORIGIN OF SPECIES BY MEANS OF NATURAL SELECTION, OR THE PRESERVATION OF FAVOURED RACES IN THE STRUGGLE FOR LIFE,* LAYS OUT ITS MAIN IDEA: THAT LIFE-FORMS DEVELOP THROUGH A PROCESS OF ADAPTATION THAT LEADS TO EVOLUTION OVER TIME.

Evolution Is True, "A theory is much more than just a speculation about how things are: it is a well-thought-out group of propositions meant to explain the facts about the real world.... [T]he theory of evolution ... is an extensively documented set of principles."

In other words, a theory is far more than a scientific hypothesis, which is an educated guess at the way nature works. A theory is an explanation of facts that is backed up by a large body of evidence. A theory is a powerful set of ideas that has become stronger and more broadly accepted by withstanding challenges.

This is the story about how a hypothesis became not merely a theory but the foundation of an entire science. And it is the story of a remarkable individual, Charles Darwin, who saw the importance of that theory and risked controversy and even ridicule to bring it to light.

A YOUNG NATURALIST

February 12, 1809, was a remarkable day. On opposite sides of the Atlantic Ocean, two baby boys were born who would change the world. Abraham Lincoln became president of the United States during the darkest days of the Civil War, and, in England, Charles Robert Darwin grew up to become the scientist who wrote the most revolutionary book in the history of biology.

Young "Bobby," as Charles was called, was the grandson of two important families. His father's father, Erasmus Darwin, was a poet and physician. His mother was the daughter of Josiah Wedgwood, founder of Wedgwood pottery. Life in early nineteenth-century England was fairly quiet for Charles, that is, until his mother died when he was just eight years old.

CHARLES DARWIN WAS THE SON OF TWO IMPORTANT FAMILIES. HIS GRANDFATHERS WERE POET AND PHYSICIAN ERASMUS DARWIN AND JOSIAH WEDGWOOD, FOUNDER OF WEDGWOOD POTTERY. HIS PRIVILEGED UPBRINGING IS CLEAR IN THIS 1816 PORTRAIT WITH HIS YOUNGER SISTER CATHERINE BY ENGLISH PAINTER ELLEN SHARPLES. EVEN AT THIS EARLY AGE, HE WAS DEVELOPING AN INTEREST IN COLLECTING OBJECTS FROM THE NATURAL WORLD.

Charles and his older brother, named Erasmus after his grandfather, attended Reverend Samuel Butler's Shrewsbury School. Charles showed a great interest in learning. But rather than learning Greek and Latin like most students of the time, he was taken with the English poetry of William Wordsworth and Lord Byron.

When Charles was a teenager, science began to captivate him, so much so, in fact, that he and Erasmus built a chemistry lab in a garden shed. Charles enjoyed chemistry. His friends later nicknamed him "Gas."

A DIFFERENT PATH

Robert Darwin was a physician who had high expectations for his sons. Because Charles showed little interest in the classics, Robert felt that Charles's school was a waste of time and in 1825 withdrew him. Thinking that his son had no direction in life and not realizing that Charles had developed an interest in science, particularly animal life, he decided that Charles was to become a doctor. With that aim in mind, he sent his son to join Erasmus at the University of Edinburgh in Scotland.

But Charles had other ideas for himself. He disliked the sight of blood and found his medical studies tedious. He did, though, like studying the art of taxidermy, which he learned there from a former slave from South America named John Edmonstone. During his second year in Scotland, Darwin joined the Plinian Society, a club for naturalists. He was taken with their intellectual debates, which exposed him to ideas of how humans were created by gradual changes in form over time. These ideas were different from the biblical teaching that God created the universe, on which he had been raised.

RESHAPING IDEAS

While at the university, Darwin met Robert Grant, a zoologist. The two became close friends. At this time, Darwin began collecting fossils and learning more about animal life. Grant has been credited as being the first person to interest Darwin in the ideas that would lead to his theory of evolution. Instead of believing that species are as God had created them, many scientists favored the idea that organisms change over time, adapting to survive in their environments. Of particular interest to Darwin were the writings of French zoologist Jean-Baptiste Lamarck (1744–1829).

After two years, Darwin, quite sick of the sight of blood, quit medical school and went to London. He joined his uncle Josiah Wedgwood II for a trip to Paris. As Charles vacationed, Robert Darwin fretted over his son's dropping out of medical school and becoming an idle gentleman. This time he made plans for Charles to study for the ministry. He enrolled him at Christ's College at the University of Cambridge.

During the summer, before school began, Charles fell in love with Fanny Owen. Fanny was the sister of one of his best friends. They spent many hours together, talking, riding horses, and playing cards. Charles also spent that fall preparing for Cambridge, completing studies that were required before admission. He began classes at the end of 1827.

But rather than study the Bible, Charles developed a fondness for collecting beetles. This predilection led to a further interest in the field of naturalism. He attended lectures on botany given by the Reverend John Stevens Henslow (1796–1861). Darwin saw this as a possible career path, and he pursued his studies with enthusiasm.

DARWIN, FITZROY, AND THE WINDS OF CHANGE

In the history of exploration, few voyages are more famous than that of the British naval vessel *Beagle* between 1831 and 1836. Though it has become legendary for the discoveries by its young naturalist, Charles Darwin, its goal was to map and explore South America. Darwin's presence on board was an afterthought.

The ship's captain, Robert FitzRoy (1805–1865), had asked permission from Admiral Francis Beaufort (1774–1857, after whom the most commonly used wind scale is named), to bring on board a companion. FitzRoy wanted someone who would be his intellectual and social equal and who also might serve as naturalist.

The budget did not allow for another crew member, so the person selected had to be a gentleman with means to pay for his own meals. Beaufort, through his connections, recommended Darwin. The voyage shaped Darwin as a scientist in many important ways, not the least of which was his development as a meticulous record-keeper, a skill he learned from FitzRoy.

DARWIN'S FIRST JOB AS A NATURALIST WAS ON THE BRITISH NAVAL VESSEL *BEAGLE*, WHICH SET OFF TO MAP SOUTH AMERICA IN 1831. ITS CAPTAIN, ROBERT FITZROY (*SHOWN HERE*), REQUESTED PERMISSION TO BRING ON A COMPANION WHO WOULD BE HIS SOCIAL AND INTELLECTUAL EQUAL AND SERVE AS AN UNPAID NATURALIST. DARWIN MORE THAN FULFILLED THAT ROLE; HIS COLLECTIONS AND DISCOVERIES ON THE VOYAGE FOREVER CHANGED THE LIFE SCIENCES.

FROM BEETLES TO THE *BEAGLE*

By 1829, it was clear that Darwin was not interested in joining the clergy. Darwin spent his spring break with Reverend Frederick Hope (1797–1862), a noted entomologist. Reverend Hope provided more samples for Charles's growing beetle collection.

Within the year, it became apparent that Darwin preferred his study of beetles to that of his romance with Fanny. They broke up the following spring. Darwin became a regular at Professor Henslow's Friday night dinner parties, and botany became his passion.

EVEN WHILE STUDYING FOR THE MINISTRY IN 1829, DARWIN'S INTEREST IN THE NATURAL WORLD DOMINATED HIS THINKING. HE SPENT HIS SPRING BREAK THAT YEAR WITH NOTED ENTOMOLOGIST REVEREND FREDERICK HOPE, WHO PROVIDED SAMPLES FOR THE BUDDING NATURALIST'S GROWING BEETLE COLLECTION.

Despite his general lack of interest in the course-work, Darwin passed his final exams in January 1831, placing tenth in his class. Finished with school, Darwin was ready to become a countryside clergyman while indulging in his scientific interests in his personal time. Henslow, though, suggested to Darwin that he travel first, words that would prove fateful.

That spring, Darwin prepared for a summer trip to the Canary Islands, off the coast of Spain, while attending geology lectures from Professor Adam Sedgwick (1785–1873). Geology had bored Darwin in the past, but Sedgwick brought the topic to life and fully engaged Darwin's attention.

Rather than go alone, Darwin asked his friend Marmaduke Ramsay (ca. 1796–1831) to accompany him on the journey. In the final weeks before departure, Sedgwick lectured about geology so Darwin would know what he was observing on the islands. Darwin also went with Sedgwick on a geologic field trip to the mountains of Wales in early August. The trip was the only time that Darwin had received formal scientific training in the field. But in mid-August Darwin's plans came to a crashing halt—Ramsay suddenly died. Darwin was stricken with grief and could not travel to the Canary Islands. Then Professor Henslow and Reverend George Peacock wrote to Darwin and told him about an opening for a naturalist on a ship—the H.M.S. *Beagle*.

A few months later, Charles Darwin was bound for South America—and for history.

CHAPTER TWO

A TIME OF EVOLVING THOUGHT

Although science is always about facts and evidence, it is always influenced by the political and social environment of its time. That was certainly true of Darwin's decisions about how to present his discoveries.

Darwin lived in a century of great discovery and technological advancement. He was born as scientists were just beginning to understand the atomic and chemical nature of matter, soon followed by in-depth studies of electricity and magnetism. The Industrial Revolution had begun to transform society. Many scientists toured and lectured before public audiences. Darwin was exposed to many of these throughout his formative years.

CREATIONISM

Still, much of the world's population turned to religion instead of science for an explanation as to how the universe and life on Earth came into being. Many people believed then, as they do today, that God created the universe and all of the creatures on Earth approximately five thousand years ago, which corresponds to a particular interpretation of the Bible.

Today, this belief is known as Young-Earth Creationism. Others accept the geological evidence that Earth is much older and that species have evolved, but still believe that the world was created by God (or by a supernatural designer) who sets guidelines for evolution. This belief is called creation science if it includes God or intelligent design (ID) otherwise. Advocates claim that creation science and ID are sciences, but most scientists, even those who believe in God, reject that claim. Science must permit any hypothesis to be tested, and that includes the existence of God or the designer.

During Darwin's time, leaders of most Christian churches called anyone who did not believe in creationism a heretic, someone who rejects Church doctrine. Many of Darwin's fellow scientists were beginning to question creationism. Why did some creatures live only in certain parts of the world? Why did some creatures, such as the recently discovered dinosaurs, become extinct? Why did certain species die out, leaving only a fossil record, while other similar species flourished?

EVOLUTION IN HISTORY

The theory of evolution was not new to Darwin. It began long ago, but it never had scientific evidence of steady change over a long time. Some ancient Greek philosophy texts that have survived the ages pondered evolution. In the many centuries since, various scientists and philosophers have touched on the subject of evolution, but they were usually lone voices, largely ignored by their peers and the people of the time. Often, people did not want to hear what the scientists had to say because they feared the religious authorities, who were often political rulers as well. They would not stand for anything that differed from the religion's belief about how the world was created, and they had the power to punish anyone who disagreed.

Still, by the eighteenth century, it had become more difficult to deny the evidence of evolutionary connection among species. For example, in 1745, the French philosopher and scientist Charles Bonnet created a linear chart showing those physical connections, with humans at the top and mold at the bottom. (Bacteria had not yet been discovered.) He was far from the only scientist to classify life-forms.

Carolus Linnaeus (1707–1778), a Swedish botanist and explorer, developed a classification system that grouped animals into categories such as Canidae (dogs) and Felidae (cats). From here, he developed theories of hybridization, showing how different species mated and produced creatures of mixed species. Linnaeus's theory of hybridization suggested that species could change over time, supporting the idea that they evolve.

Linnaeus's ideas resembled those of French naturalist Georges-Louis Leclerc, count de Buffon (1707–1788). Buffon's theory of "degeneration" states that by looking at a species today, its lineage can be traced back to a primal ancestral starting point. This ancestral starting point was one species from which all others evolved. For example, an ancient feline, or

DARWIN'S BREAKTHROUGH WAS NOT THE IDEA THAT SPECIES EVOLVED. IT WAS EXPLAINING WHAT CAUSES EVOLUTION. ONE OF THE FIRST SCIENTISTS TO DESCRIBE EVOLUTION WAS EIGHTEENTH-CENTURY SWEDISH BOTANIST AND EXPLORER CAROLUS LINNAEUS, WHO DEVELOPED A SYSTEM FOR CLASSIFYING ANIMALS AND PLANTS. HIS IDEA, WHICH TURNED OUT TO BE INCORRECT, WAS THAT DIFFERENT SPECIES MATED TO PRODUCE MIXED CREATURES.

cat, species developed into present-day lions, tigers, pumas, and house cats.

Charles Darwin's own grandfather, Erasmus Darwin, combined the theories of creationism and evolution. In his book *Zoonomia*, he put forth the notion that God designed life but that God also designed it to be self-improving. (This view is similar to intelligent design.) He said that animals were meant to grow and adapt to their changing surroundings. Erasmus so impressed and influenced Charles's thinking that, in his later years, Charles wrote a biography about his grandfather.

A WORLD TO EXPLORE

During this time of discovery, the H.M.S. *Beagle*

ERASMUS DARWIN WROTE A BOOK CALLED *ZOONOMIA* THAT PROPOSED THAT GOD HAD DESIGNED LIFE TO BE SELF-IMPROVING. HIS IDEA WAS THAT EVOLUTION TOOK PLACE BECAUSE ANIMALS WERE MEANT TO GROW AND ADAPT TO THEIR SURROUNDINGS. ALTHOUGH HE EXPLICITLY INCLUDED GOD AND DID NOT DESCRIBE NATURAL SELECTION, HIS IDEAS WERE A MAJOR INFLUENCE ON HIS GRANDSON CHARLES.

was being prepared as one of several ships scheduled to map South America. When Captain FitzRoy offered Charles an unpaid spot on the crew as naturalist, Charles's father, Robert, refused to give his adult son the funds to support himself on the voyage. Fortunately, Charles's uncle Josiah Wedgwood II stepped in and persuaded Robert to allow Charles to join the excursion, which was scheduled to depart from Plymouth Harbor in September 1831. The departure date was gradually moved back, and the ship finally set sail on Tuesday, December 27.

DARWIN'S
TREASURE TROVE

*D*arwin certainly expected to learn a lot about nature on his upcoming voyage, but he never could have anticipated the number of samples he would collect. Nor could he have expected the importance of the research it would inspire. His voyage on the *Beagle* would carry him around the world and produce a treasure trove of samples large enough to last a lifetime of research. His collection would inspire a new theory that would transform biological science forever.

CROSSING THE ATLANTIC

In early December 1831, a few weeks before it set sail, Darwin boarded the H.M.S. *Beagle* for the first time. He prepared his small cabin, located above the captain's quarters. The space was 9 feet (2.7

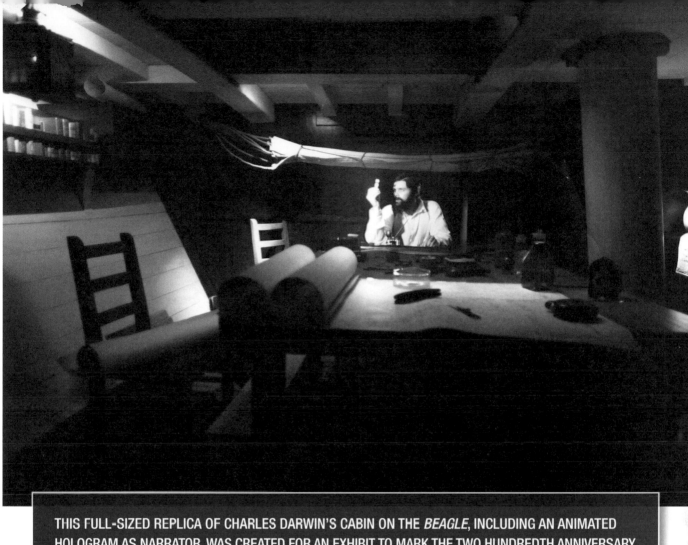

THIS FULL-SIZED REPLICA OF CHARLES DARWIN'S CABIN ON THE *BEAGLE*, INCLUDING AN ANIMATED HOLOGRAM AS NARRATOR, WAS CREATED FOR AN EXHIBIT TO MARK THE TWO HUNDREDTH ANNIVERSARY OF DARWIN'S BIRTH IN A MUSEUM IN DARWIN'S FORMER HOME, DOWN HOUSE IN KENT, ENGLAND. NOTE THE SHIP'S MAST RUNNING THROUGH IT ON THE RIGHT.

meters) wide by 11 feet (3.4 m) long and only 5 feet (1.5 m) high. Part of the cabin was taken up with one of the masts rising through it.

Darwin boarded early because he wanted to adjust to shipboard life prior to departure. When the ship finally weighed anchor at around 11:00 AM on December 27 and headed southwest toward Madeira Island, Darwin was almost immediately seasick, a condition

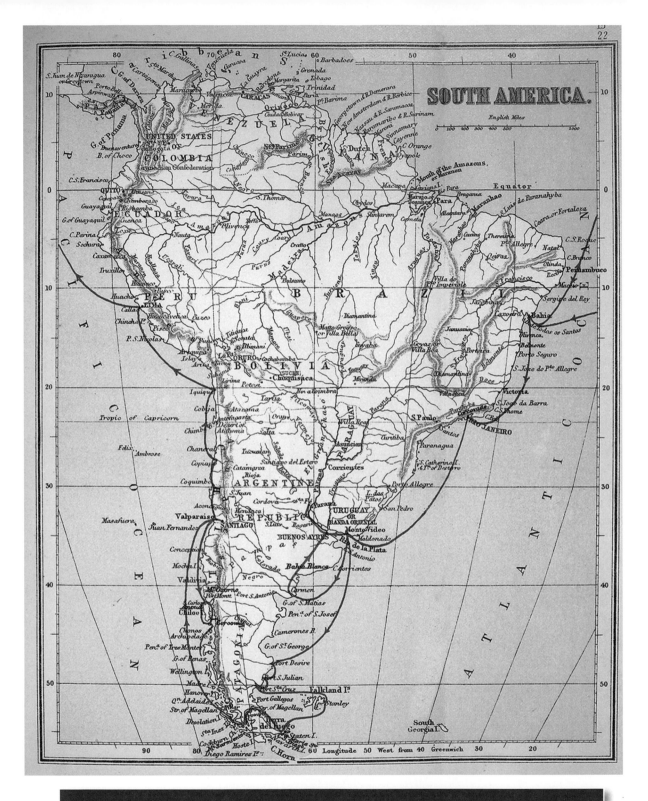

South America.

THIS MAP OF SOUTH AMERICA IS FROM THE APPENDIX OF THE 1890 EDITION OF DARWIN'S BOOK *THE VOYAGE OF HMS BEAGLE*. BESIDES SHOWING PLACES THAT THE *BEAGLE* DOCKED ON THE COAST, IT SHOWS DARWIN'S EXPLORATIONS ON RIVERS AND INLAND. THE GALÁPAGOS ISLANDS, WHERE HE WOULD MAKE HIS MOST FAMOUS OBSERVATIONS OF THE EVOLUTION OF FINCHES, ARE JUST BEYOND THE WESTERN (LEFT) EDGE OF THE MAP.

that never truly left him during the entire voyage. The *Beagle* touched the shore of the island of Madeira to confirm its position but did not stop. It continued toward its first port of call at the town of Santa Cruz on Tenerife in the Canary Islands on January 6, 1832. Because England had been experiencing the disease of cholera, the members of the crew were not allowed to get off the ship without a period of quarantine. Rather than waiting, FitzRoy insisted that they sail on.

Darwin finally set foot on foreign soil on January 16, 1832. The crew stopped at the Cape Verde Islands, off the western coast of Africa. Almost immediately Darwin made a discovery that combined his various interests in geology and naturalism. It made him think. He found a band of fossil shells 45 feet (13.7 m) above sea level. This finding made him question how the fossils rose so high.

BRAZIL TO TIERRA DEL FUEGO

The *Beagle* remained in port for twenty-three days. It then set out for Brazil, arriving in the port city of São Salvador at the end of February 1832. Darwin took solitary walks, marveling at the rich foliage and exotic animal life. The *Beagle* remained in Brazil through the spring of that year. The ship traveled along the coast, stopping in many ports of call. At each stop, Darwin continued to take long hikes and collect countless specimens.

THE *BEAGLE* MADE ITS FIRST LANDFALL IN SOUTH AMERICA ON THE COAST OF BRAZIL AT THE END OF FEBRUARY 1832. OVER THE NEXT SEVERAL MONTHS, THE SHIP EXPLORED THE ATLANTIC COAST, EVENTUALLY REACHING THE SOUTHERNMOST REGION OF THE CONTINENT KNOWN AS TIERRA DEL FUEGO BY THE END OF THE YEAR. THIS PAINTING BY CONRAD MARTENS, AN ARTIST HIRED BY CAPTAIN FITZROY TO DOCUMENT THE CREW'S EXPLORATION, SHOWS THE *BEAGLE* IN THAT REGION'S MURRAY NARROWS.

In August, Darwin sent a load of specimens, along with detailed notes describing them, back to Henslow in Cambridge for safekeeping. By then, Darwin had earned the nickname of "Flycatcher" or "Philosopher" from the ship's crew because of his large collection of specimens. FitzRoy called the growing collection junk, but each item was precious to Darwin.

Next, the ship sailed to Patagonia, a large southern region in South America that is now part of Argentina. There, Darwin collected fossils, bits of bone, and feathers, among other things he had never before encountered. He struggled to accurately record their features in the hopes that more experienced naturalists could later help to identify them.

Darwin sent a second shipment to Henslow in November, before the ship headed for Tierra del Fuego, on the southern tip of South America. There, Darwin first encountered native peoples who still lived in the wilderness and were considered to be savages by the Europeans.

THE FALKLANDS AND URUGUAY

In March 1833, the crew of the *Beagle* began mapping the Falkland Islands, which the British had claimed from Argentina just months earlier. Fascinated by bird and animal fossils found there, Darwin spent his time comparing those specimens with everything he had collected to date.

In May of that year, Darwin remained in Maldonado, Uruguay, while the ship headed back to Montevideo. He took a twelve-day journey into the interior, using hired gauchos, or cowboys, for help. Recognizing the need for full-time help, Darwin wrote home, asking his father to provide money so he could hire a servant. When permission arrived, Darwin asked the ship's odd job man, Syms Covington (1816–1861), to take the role.

After sending a third shipment of specimens to Henslow in July 1833, Darwin was ready to see new locales. Darwin later learned that Henslow had taken to reading his letters aloud to the Philosophical Society of Cambridge. This attention earned Darwin an early reputation there as being an excellent naturalist and observer.

In August, the *Beagle* arrived at the Río Negro. Darwin traveled inland again, ultimately arriving at Bahia Blanca. At one point, he found a fossil of an animal he had never seen before. The fossil was under a layer of white shells similar to the ones he first saw months earlier. He puzzled over this new discovery, wondering how these shells, which he first found in Cape Verde 45 feet (13.7 m) above sea level, also turned up at ground level in South America. Also, what animal made these fossils? As the group left Montevideo in December, Darwin sent his fourth shipment to England. In Uruguay, he continued to collect fossils, consumed by this new passion.

AROUND THE HORN TO CHILE

On Darwin's twenty-fifth birthday, Captain FitzRoy named the highest mountain in the Tierra del Fuego region Mount Darwin, in Charles's honor. By April 1834, the ship had sailed around Cape Horn at the most southern tip of South America, leaving the Atlantic Ocean and entering the Pacific Ocean.

Darwin became quite ill that summer. Historians believe that this illness, the first of many noteworthy

MEASURING LONGITUDE

Why was Captain FitzRoy so concerned about measuring longitude? In the early days of exploration, it was very difficult to determine a ship's exact position once it was out of sight of land. Its latitude, the position north or south of the equator, could be determined by the stars at night or the angle of the sun above the horizon at its high point for the day.

That high point also told sailors when it was "local noon," or midday as measured by the sun's position. If they knew what time it was in England (or some other location), knowing local noon would tell them how far east or west they had traveled in latitude. If it was 2:00 PM in Greenwich, England, (latitude zero) when the sun passed its high point for them, they knew they had traveled as far west as the sun does in two hours. Because the sun travels a full circle in twenty-four hours, that would mean they had traveled one twelfth of the way around the world, or 30 degrees west in longitude.

The only way for FitzRoy to be sure of the time in Greenwich was to have a very accurate clock, called a chronometer, on board. Whenever he landed at a port where he knew the latitude, he would reset and adjust his chronometer to keep it running as accurately as possible.

illnesses, turned him into a recluse in later years. (Based on the recorded symptoms, doctors today suspect he was infected during the voyage with Chagas' disease, a form of sleeping sickness.) He was deemed unfit to return to the ship until late October.

Finally in November, Darwin returned to the *Beagle*. It headed for the Chronos Archipelago off southern Chile. Darwin found wild potatoes growing there, which amused him. He didn't expect to see potatoes growing so far away from home.

In February 1835, while studying farther north in Valdivia, Chile, Darwin experienced his first earthquake. Although the quake did not do much damage to Valdivia, some surrounding areas were hit hard. As the ship and crew traveled north along the Atlantic coast, they saw the devastation caused by the quake.

The earth was giving Darwin clues. In March, when Darwin spent a few days on the island of Quiriquina and in what was left of the devastated port town of Concepción, he noted that some of the land had risen several feet as a result of the earthquake. He concluded that much of South America was rising because of such activity. This observation confirmed British geologist Charles Lyell's (1797–1875) belief that landmasses rose over time.

Darwin had been introduced to Lyell's work years earlier by someone who suggested he would get a laugh out of it. Instead, he was fascinated, and now here in Chile he was finding evidence to support Lyell's ideas. The recent earthquake's effects on the land plus the Cape Verde fossils gave credence to Lyell's theories. Landmasses did rise, and Earth was probably far older than previously thought.

THE WONDERS OF THE GALÁPAGOS AND TAHITI

The *Beagle* continued on its northward track along the Pacific Coast of South America, arriving in Callao, near Lima, Peru, in July to take on provisions for the next leg of the journey. It then headed northwest to the Galápagos Islands, 500 miles (800 km) west of Ecuador. On September 15, 1835, members of the crew spotted the next port of call—the Galápagos's Chatham Island.

Darwin was intrigued by the black lava covering the island's shoreline. He had never seen anything like it before. It was the first of many surprises he would find in the Galápagos. For the next couple of weeks, the crew explored several islands. In October, Darwin and several others stayed on James Island for one week to continue their research. Darwin thought he collected an impressive number of specimens there, including several finches, which would later be extremely important in formulating his ideas about natural selection. Still, his Galápagos collection was dwarfed in number by what he discovered when the ship arrived in the Polynesian island of Tahiti that November. He was amazed by the variety and quantity of lush vegetation, which was unlike anything he had witnessed since leaving Europe.

AUSTRALIA AND NEW ZEALAND

By early 1836, the *Beagle* found its way to Australia and New Zealand. Although neither of these

IN EARLY 1836, THE *BEAGLE* REACHED THE COCOS (OR KEELING) ISLANDS NORTHWEST OF AUSTRALIA, WHICH DARWIN NOTICED WERE ALMOST ENTIRELY MADE OF CORAL. A FLAT CORAL REEF WAS ALSO VISIBLE JUST BELOW THE OCEAN'S SURFACE NEARBY. HE SPECULATED THAT SEA LEVEL MUST HAVE LOWERED OVER TIME, EXPOSING REEFS THAT HAD ONCE BEEN COMPLETELY SUBMERGED.

ports impressed Darwin as much as South America, he did take time to study the geology of both lands to compare against his notes from the *Beagle's* other stops. For example, when they stopped at the Cocos (or Keeling) Islands northwest of Australia in the Indian Ocean in April, Darwin noted that the islands were almost entirely made of coral. He speculated that they must have been part of a huge coral reef that was once submerged but was now exposed due to the lowering sea level.

By summer, the crew was ready to go home. They made their way around the southern tip of Africa. To check his ability to measure longitude, FitzRoy then crossed the Atlantic to Bahia in Brazil, before heading back to Europe. After four years, nine months, and five days, the *Beagle* arrived in Falmouth, England, on October 2.

Darwin's voyage of a lifetime was completed. But his work was only beginning. His collections of fossils and samples of animal and plant life were full of surprises waiting to be revealed and mysteries waiting to be solved.

CHAPTER FOUR

DISCOVERIES AND QUESTIONS

Now twenty-seven years old, Darwin arrived home to begin his life as an independent adult and a scientist with a trove of research material to keep him busy.

Almost immediately, he set about corresponding with fellow scientists, making plans to analyze the materials he had gathered. Darwin encountered problems at various museums because they were flooded with fossils and other material from the British colonies around the world. The backlog was going to present a serious challenge to Darwin in getting his findings into those collections.

At his uncle's suggestion, Darwin began organizing his records and thought about writing a book. After spending time in seclusion, cataloging his fossils, Darwin finally made an eagerly anticipated public appearance thanks to Reverend John Stevens Henslow's reading his letters to the

DARWIN, SHOWN HERE IN AN 1840 PAINTING, RETURNED HOME TO DISCOVER THAT HIS LETTERS FROM HIS VOYAGE ON THE *BEAGLE* HAD MADE HIM FAMOUS. HE WAS SOUGHT AFTER AS A SPEAKER, AND HIS WRITINGS ABOUT HIS ADVENTURES AND DISCOVERIES WERE READ EAGERLY. BUT HE KNEW THAT HE HAD BARELY BEGUN TO ANALYZE HIS COLLECTION OF ANIMAL AND PLANT SPECIMENS AND HIS GEOLOGICAL SAMPLES.

Philosophical Society of Cambridge. He presented a paper to the Royal Geological Society on January 4, 1837, discussing his belief that South America rose through geologic movements over time and how the local species adapted to their new surroundings.

FOSSILS AND FINCHES

Darwin continued his research through the spring, still aided by Syms Covington. Slowly, he came to the realization that the various fossils in his possession suggested that species develop differences from one another over time. And he realized that his collection might also include evidence of the same kind of development in species that were still living and flourishing.

For example, the birds that Darwin collected on the Galápagos included what appeared to be distinct species of finches. Darwin wanted to understand how these distinct forms came to be. He identified fourteen distinct finch species spread over the dozen small islands of the Galápagos. To Darwin, the best explanation is that a single species of finch from the mainland was the ancestor of all of the species he had identified. Because the mainland was about 500 miles (800 km) away, the number of finches reaching each island would be small. Over time, these finches evolved into new species, each of which was well adapted to the specific environments of its own island.

DARWIN NOTICED DIFFERENCES BETWEEN SPECIES OF FINCH FOUND ON DIFFERENT ISLANDS IN THE GALÁPAGOS, SUCH AS THE ONE SHOWN HERE. THAT LED HIM TO REALIZE THE ROLE THAT NATURAL SELECTION PLAYS IN THE DEVELOPMENT OF SPECIES. TODAY, THE BIRDS ARE OFTEN REFERRED TO AS DARWIN'S FINCHES BECAUSE OF THEIR IMPORTANCE IN THE HISTORY OF SCIENCE.

As Darwin worked on his own studies, he continued to stay abreast of other scientific developments. When he read about monkey fossils found in Africa, Darwin noted only to himself the idea that humans may trace their origins to these animals because of the physical similarities between the species. However, he was learning during this time that despite the fascinating scientific discoveries being made, most scientists still held tightly to their religious beliefs of creation. It would be risky to suggest that humans, apes, and monkeys were related without powerful evidence.

NOTES AND QUESTIONS

By the summer of 1837, Darwin was compiling his notes into book form. He and FitzRoy were collaborating on a multivolume work titled *Narrative of the Surveying Voyages of His Majesty's Ships* Adventure and Beagle, *between the years 1826 and 1836.* Darwin's portion, called *Journal and Remarks, 1832–1836,* was to be the third volume of the set. At this time, however, Darwin was suffering from bouts of his illness. Nevertheless, he completed his portion of the manuscript.

Captain FitzRoy was late with his contribution, which delayed publication until 1839. A few weeks after the multivolume set was published in 1839, Darwin's book was reissued under the title of *Journal of Researches into the Geology and Natural History of the Various Countries Visited by H.M.S.* Beagle, *under the Command of Captain FitzRoy, R.N., from 1832 to 1836.* The work is most commonly referred to as the *Journal of Researches* or sometimes the *Voyage of the* Beagle.

While waiting for FitzRoy to complete his manuscript, Darwin also started a new notebook in the summer of 1837, where he posed several questions to himself:

- What evidence is there that species go through the process of transmutation?
- How do species adapt to a changing environment?
- How are new species formed?
- Why are there similarities between different species?

In his notes, Darwin began to theorize how species changed over time because of changes in their environment. He called that process transmutation. He diagrammed the way a particular species evolved over time. One of the challenges that Darwin faced was determining how animals

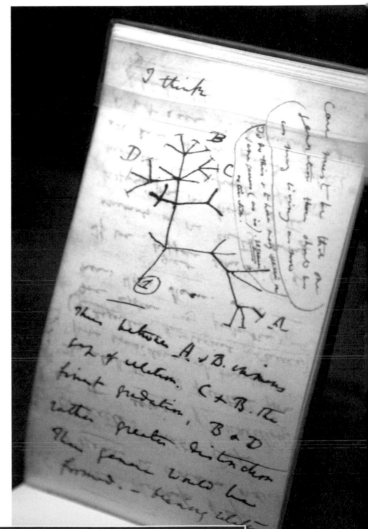

IN THIS NOTEBOOK, DARWIN'S WORDS "I THINK" ARE FOLLOWED WITH A DIAGRAM THAT SHOWS THE WAY HE THOUGHT SPECIES MIGHT EVOLVE OVER TIME. HE BEGAN TO CONSIDER THE RELATIONSHIP BETWEEN EVOLUTION AND A SPECIES' CHANGING ENVIRONMENT. THAT EVENTUALLY LED HIM TO THE IDEA THAT NATURAL SELECTION WAS THE DRIVING FORCE FOR EVOLUTION, WHICH HE AT FIRST CALLED TRANSMUTATION.

moved from one place to another like the finches that migrated from the mainland to the Galápagos Islands. These questions were the basis of Darwin's research through the fall and into the winter of 1838.

At the end of winter, Darwin began another notebook. Known as his C Notebook, Darwin filled the pages with his theories about transmutation. By this time, Darwin had written to experts in various fields, from dog breeders to veterinarians, asking very specific questions about crossbreeding—the mating of different species or breeds to produce a new form of life. The answers he received from the experts helped him to refine his theories about how crossbreeding took place in the wild. Darwin was careful not to announce his theories before they were fully developed. After all, there was no value in risking public censure until his theories could be defended successfully.

It was becoming clear to Darwin that species adapted to their environments. Those species that changed to become more fit for their environments survived, and even thrived. Other species that did not adapt died out. Darwin began to question not only why this happened but also how this happened.

He found a possible answer to these questions in an old pamphlet, *The Art of Improving the Breeds of Domestic Animals* by Sir John Sebright (1767–1846), published the year Darwin was born. Sebright suggested that the weaker species do not live long enough to pass on their traits to the next generation. Darwin thought about these ideas with regard to his finches. It occurred to him that the finches would have multiplied to the point where the resources on the mainland would have been fully used. Only the finches that were most fit would have been able to survive the several-hundred-mile journey to the islands and thrive in their new environment.

DARWIN'S MARRIAGE TO EMMA

By the summer of 1838, Darwin felt bold enough to share his radical ideas with his father, who took them in stride. During the last several months, Darwin had struck up a close relationship with his cousin, Emma Wedgwood. However, when he brought up the subject of marrying Emma, his father warned that she came

AS DARWIN GREW MORE CONFIDENT IN HIS SCIENCE, HE ALSO GREW CLOSER TO HIS COUSIN EMMA WEDGWOOD, SHOWN HERE IN THIS 1840 PAINTING. BY 1838, THEY WERE FALLING DEEPLY IN LOVE, AND HE PROPOSED MARRIAGE ON NOVEMBER 11 OF THAT YEAR. THEY MARRIED THE FOLLOWING JANUARY 29 AND HAD TEN CHILDREN, SEVEN OF WHOM LIVED INTO ADULTHOOD.

from a very strict family—Emma's family would never consider Darwin's theories as anything but heresy. Deeply in love, Darwin ignored his father's advice. He mentioned some of his notions regarding religion and nature to Emma, who was surprisingly tolerant. Emma often pleaded with Darwin not to approach religion the way he did science. She was unhappy that their views about religion and God were so contrary, but she deeply admired him.

Darwin juggled numerous tasks, from writing books to studying more fossils to watching orangutans and other apes at the London Zoo. As he worked, he also continued his relationship with Emma. He proposed to her on November 11, 1838, and they married the following January 29. The marriage was well received by both sides of the family. The couple settled into a home in London, which was already filled with scientific material.

Arrangements were made for Darwin to receive a handsome annual sum of money from his father. These funds would not only allow the couple to live comfortably but also enabled Darwin to retain the services of a research assistant, first Covington and then Joseph Parslow (1812–1898).

THE PUBLICATION OF *BEAGLE*

During the spring of 1839, Darwin continued his research on crossbreeding. He proceeded to ask various

DARWIN'S CHILDREN

Charles and Emma Darwin's marriage produced ten children, beginning with the birth of William Erasmus on December 27, 1839. William grew up to be a banker. Anne Elizabeth ("Annie"), born in 1841, died of tuberculosis at the age of ten. Darwin was deeply affected by her death. Even though he was qualified to serve as a minister, he was unable to accept Annie's death as part of God's plan. His Christian beliefs were shaken to the core by that painful loss. Daughter Mary Eleanor lived less than a month after her birth in 1842. Henrietta Emma ("Etty"), born in 1843, lived to adulthood and edited and published Emma's personal letters. George Howard, born in 1845, became an astronomer and mathematician. Elizabeth, who was born two years later, lived to adulthood and never married. Born in 1848, the Darwins' son Francis became a botanist. His brother Leonard, born in 1850, was trained as a military engineer. Horace, born in 1851, made scientific instruments and formed a company. The Darwin's last child, Charles Waring, born in 1851, lived less than nineteen months.

experts, including farmers, questions about how they crossbred their animals. He filled pages with his correspondences. In May, the multivolume collection he wrote with Captain FitzRoy, *Narrative of the Surveying Voyages of His Majesty's Ships* Adventure *and* Beagle, was finally released to the public.

In August, the third volume of the *Narrative* series was reissued under the title *Journal of Researches into the Geology and Natural History of the Various Countries Visited by H.M.S.* Beagle. Though Darwin was already recognized as a scientist, that account of the travels and discoveries on the *Beagle* expedition brought him broad public acclaim as well. Over the years, the *Journal of Researches* was released in many different editions, under many different titles. The most well known version was published in 1845, which included more details on his findings. It remains in print today.

In June 1839, Darwin completed his N Notebook, comprising his research on transmutation. Still worried about how the public would react to his developing theories, Darwin held off on publishing his findings. He would continue to hold off for another twenty years.

SKELETON OF THE MYLODON DARWINII.
Page 106

AMONG DARWIN'S MANY DISCOVERIES ON THE *BEAGLE*'S ROUND-THE-WORLD VOYAGE WERE BONES AND SKELETONS OF EXTINCT ANIMALS, SUCH AS THIS GROUND SLOTH THAT CAME TO BE KNOWN AS *MYLODON DARWINII.* THIS SKETCH OF *MYLODON* APPEARED IN THE 1890 EDITION OF THE BOOK *JOURNAL OF RESEARCHES* BY CHARLES DARWIN.

THE THEORY OF EVOLUTION EVOLVES

By 1842, Darwin accepted his grandfather's theory that God set the universe into motion only to let species evolve on their own, combining both the creationist theory with that of evolution. During the next several months, he spent time refining the outline for his next book, which was to contain his theories. His health also began to improve. After a three-year period of illness, he soon completed his notes from the *Beagle* voyage.

Darwin's scientific research also continued. He had befriended the botanist and explorer Joseph Dalton Hooker (1817–1911) and entrusted him to examine the Tierra del Fuego plant samples. Darwin also shared his developing ideas about evolution and was thrilled by Hooker's enthusiastic reception.

Soon after meeting Hooker, Darwin completed a 189-page outline on the transmutation theory. The draft proposed that animal and plant species remained unchanged until their natural environments changed. A catastrophic event such as an earthquake could trigger adaptive changes in species. The organisms that displayed these new adaptations were the ones to survive and thrive over time, leading to a new species. To Darwin, this theory explained the variation that science observed among species.

In September, the manuscript was 231 pages in length. Darwin finally had the confidence now to share it with Emma. To his delight, rather than calling the work heretical, Emma calmly accepted many of his thoughts and made constructive suggestions.

A month later, Darwin was surprised to learn that Robert Chambers's book *Vestiges of the Natural History of Creation*, discussing transmutation, had been published. Although critics bashed Chambers for ignoring Christian beliefs about creation, the book became a best seller. Now competing against Chambers, Darwin began seeking other supporters beyond Hooker. And he began to refine his theory by going beyond the fact of transmutation to the processes that made it work. He was keenly aware that his explanation of transmutation to the scientific community and the general public needed to be both as clear as possible and well-supported by evidence to avoid protests of heresy.

As winter turned to spring in 1855, Darwin thought he had figured out how plant life spread from place to place, even across oceans. To simulate a seed being carried across a body of water, he soaked seeds for several months. He then planted them and to his surprise, the seeds took root and bloomed. Darwin then began asking colleagues around the world for verification that certain plants could be found along the shores around the world.

As the results of Darwin's queries to his colleagues came back, Darwin was delighted to learn how far many of these plants had traveled. He also asked officials of the British survey fleet whether any crew members aboard ships had spotted animals on natural formations such as small islands and icebergs. Their accounts, too, confirmed Darwin's theory that species could travel from one area to another, to interbreed and create hybrid species. Darwin was delighted.

For more practical research, Darwin began breeding pigeons of different species. This way, he could watch the results of crossbreeding and make his own observations instead of relying on those he made in the wild. Shortly thereafter, Charles Lyell shared with Darwin a paper titled "On the Law which has Regulated the Introduction of New Species," by naturalist Alfred Russel Wallace (1823–1913). (Sometimes referred to as

IN 1855 AND 1858, DARWIN'S FELLOW BRITISH NATURALIST ALFRED RUSSEL WALLACE PUBLISHED PAPERS WITH HIS IDEAS ABOUT HOW NEW SPECIES EVOLVED. THOSE WERE MUCH SHORTER AND LESS COMPLETE THAN DARWIN'S STILL UNPUBLISHED MANUSCRIPT, AND DARWIN DISAGREED WITH PARTS OF THEM. BUT WALLACE'S PUBLICATIONS SPURRED DARWIN TO COMPLETE AND PUBLISH *ON THE ORIGIN OF SPECIES* IN 1859.

the "Sarawak Law" paper, Wallace wrote the paper in 1855 while he was in the Sarawak region of Borneo, and it was the first paper in which he mentioned his arguments for the evolutionary process.) Wallace's theories were remarkably similar to Darwin's.

Darwin was inspired. On May 14, 1856, he began writing his own essay on natural selection, which he explained as the ability of one species to survive over another because it is better adapted to its environment. The intended short essay consumed the next year of his life, until exhaustion forced him to rest in the spring of 1857. He was back to work that summer, and it was clear the essay was going to become a book.

DARWIN'S LANDMARK BOOK

As Darwin continued writing the essay into 1857, he saw it grow longer and longer. He became worried that no one would read such a huge volume. He also admitted to colleagues that he had been at work on the notion of human origins for twenty years but would not discuss it beyond fellow naturalists.

In June 1858, Wallace wrote another paper, "On the Tendency of Species to Form Varieties; and on the Perpetuation of Varieties and Species by Natural Means of Selection." This time Wallace's theories were far closer to those of Darwin. Needless to say, Darwin was quite concerned that his own theories might not be original. He comforted himself by believing that the ideas expressed in Wallace's paper were not identical to his. Also, as in the first paper, Darwin disagreed with several of Wallace's findings.

Weeks later, on July 1, Wallace's paper, as well as extracts from Darwin's own writings, were read at London's Linnean Society, a group of scholars who exchanged ideas in the field of biology. This reading was the first time Darwin's theories were presented to the public. Such revolutionary ideas as transmutation and natural selection, from not one but two naturalists, brought extreme reactions, exactly as Darwin predicted.

Immediately afterward, Darwin took his family on vacation. He spent time trying to find ways to shorten his manuscript, which he finally shared with Hooker in March 1859. In the months that followed, Darwin fell ill again, but he managed to proofread the manuscript. He sent the completed text to John Murray Publishers in October. He then took the family to Yorkshire, where he sought treatment for his illness. This break allowed him to stay away from London and the reaction the book's publication was sure to elicit.

On November 24, Darwin received a finished copy of the book. The title was long, *On the Origin of Species by Means of Natural Selection, or the Preservation*

IN EARLY 1859, DARWIN SENT HIS PUBLISHER A BOOK PROPOSAL. THIS IS THE TITLE PAGE IN DARWIN'S HANDWRITING, WHICH READS, "AN ABSTRACT OF AN ESSAY ON THE ORIGIN OF SPECIES AND VARIETIES THROUGH NATURAL SELECTION, BY CHARLES DARWIN, M.A., FELLOW OF THE ROYAL, GEOLOGICAL, AND LINNEAN SOC., LONDON, 1859." THE PUBLISHER PERSUADED DARWIN TO SHORTEN THE TITLE TO *ON THE ORIGIN OF SPECIES THROUGH NATURAL SELECTION*.

of Favoured Races in the Struggle for Life, capturing the more than twenty years of research that had brought it into being. Darwin knew it would be controversial because it challenged the creationist religious explanations of life on Earth. He also knew that he and other scientists had the evidence to support his ideas.

It was time to put his revolutionary ideas to the challenge.

(R)EVOLUTION!

CHAPTER FIVE

From November 22, 1859, the day *On the Origin of Species* went on sale, the book's revolutionary ideas about the development of different forms of life stirred up controversy. Scientifically, the theory that species emerged by natural selection was powerfully persuasive and supported by a large body of evidence from Darwin's years of analysis. But because it challenged religious doctrine, many powerful people saw it as heresy.

The 1,250 copies available sold out quickly. Darwin started making changes for the demanded second edition. He completed those changes in early December, and John Murray more than doubled the print run to 3,000 copies. A German translation was also underway, which would bring Darwin's theories to other parts of Europe.

HOW SPECIES DEVELOPED

In *Origin*, Darwin explained the theories of natural selection and transmutation (a term that he used rather than evolution in the first five editions of the book) in the development of life on Earth. Natural selection is the theory that living organisms adapt to their changing environments in order to survive. Individual organisms whose traits make them best suited to the changes are the most "fit." That is, their traits are adaptive, and they have a better chance at survival and reproduction than other individuals of their kind. The organisms without adaptive traits are more likely to die off without reproducing.

Darwin believed that variation among living organisms in the wild was essential for adaptation and eventually led to transmutation. When the environment changes, some variant traits take on more importance for survival and the ability to reproduce. The more drastic the environmental change and the longer it lasts, the more important those variations become. As Darwin wrote in *Origins*, "Natural selection acts only by taking advantage of slight successive variations; she can never take a great and sudden leap, but must advance by short and sure, though slow steps."

"Mutation" refers to a change in physical traits, and the prefix "trans" means across. Transmutation is the process of moving across the line between an older species and a newer one by developing new traits or enhancing existing traits. For example, one individual member of the older species might be a little stronger or faster than the others. If it lives in an environment where it has to escape or fight off a predator, it is more fit to survive. Furthermore, its offspring are more likely to have those important survival traits. Thus threats to survival, whether from environmental factors or other species, drive transmutation and produce gradual changes in physical differences. Ultimately, the species is notably different from that in the past.

Darwin filled many notebooks with his observations of these variations and mutations that occurred in nature. Those that allow the organism to survive are passed along to the next generation. They then become a natural part of the species' makeup. That process is called natural selection. Darwin and others often used the phrase "survival of the fittest" in connection with natural selection. The word "fittest" in that case does not refer to an individual who may be in superb physical condition. Rather it refers to a species as a whole. A species is fit when it has adapted well to its environment and is able to reproduce successfully.

PUBLIC DEBATE

The public began debating Darwin's ideas in the pages of the local newspapers and down at the local taverns. Many people could not quite follow Darwin's writing and were confused by the forces behind natural selection. Several people tried to fit them into their own religious views.

In the scientific community, people also held opposing viewpoints, not about the science but about the theory's impact on society and politics. Friends of Darwin, such as Thomas Henry Huxley (1825–1895) and Hooker, defended his work. Huxley even coined the term "Darwinism" to describe Darwin's theories. Huxley called himself "Darwin's bulldog" as he vigorously defended Darwin's reasoning because the author himself was too ill to do it. Meanwhile, critic Richard Owen attacked Darwinism as being a danger to society because of its lack of religious values. Within six months, talk of natural selection had spread around

FROM THE DAY THAT IT WAS PUBLISHED, *ON THE ORIGIN OF SPECIES* STIRRED UP GREAT CONTROVERSY, ESPECIALLY AMONG YOUNG-EARTH CREATIONISTS WHO CONSIDERED IT RELIGIOUS HERESY. THOMAS HENRY HUXLEY, SHOWN HERE IN A JANUARY 1871 DRAWING FROM THE ENGLISH MAGAZINE *VANITY FAIR*, COINED THE TERM "DARWINISM" AND WAS AMONG DARWIN'S STAUNCHEST SUPPORTERS IN THE PUBLIC DEBATE.

the world, prompting passionate debate that went far beyond biology.

On June 30, 1860, New York University professor of chemistry John William Draper (1811–1882) spoke at the British Association for the Advancement of Science. The subject of Draper's speech was Darwin's theories as they influenced social progress. The argument was that evolution gave humans a better sense of their place in the world than religion. As a counterpoint, Bishop Samuel Wilberforce (1805–1873) was also present and gave a thirty-minute rebuttal. Wilberforce attacked Darwin's theories for going against religious beliefs.

Surprisingly, Captain FitzRoy of the *Beagle* was also in attendance. FitzRoy raised the Bible and asked people to follow its teachings and not Darwin's theory. The

WHEN SOME SCIENTISTS ARGUED THAT DARWIN'S THEORY GAVE HUMANS A BETTER SENSE OF THEIR PLACE IN THE WORLD THAN RELIGION, CHURCH LEADERS LIKE BISHOP SAMUEL WILBERFORCE ROSE IN DEFENSE OF THEIR BELIEFS. THIS JULY 1869 *VANITY FAIR* DRAWING OF WILBERFORCE CAPTURES HIS DETERMINATION EVEN THOUGH IT WAS CAPTIONED "NOT A BRAWLER."

ensuing discussion, dominated by Huxley and Wilberforce, lasted for four hours. Darwin was too ill to attend.

In January 1861, Huxley purchased the publication the *Natural History Review*, using it as a forum to promote Darwin's views on evolution and natural selection. In his very first issue, Huxley wrote about the physical similarities between humans and apes, something he theorized, like Darwin, years earlier. Huxley sent a complimentary copy to his rival Bishop Wilberforce and then set out to engage Owen in debate over the next few months. The public could not get enough of the sophisticated battle.

A NEW INTEREST BLOOMS

A year after *Origin's* publication, it remained a strong seller in England with translations in German, Dutch, and French. Meanwhile Darwin's curious mind continued to find new subjects to research.

While vacationing with his daughter Henrietta, Darwin's attention was captured by the way certain insects pollinated only certain species of flower. His research on breeding pigeons was recently completed, so he turned to the study of orchids on his return home. Once people learned of his latest interest, they sent Darwin orchid samples. The house was full of specimens once again. Darwin wrote a book on the subject titled *On the Various Contrivances by which British and Foreign*

Orchids are Fertilised by Insects, and on the Good Effects of Intercrossing. It was published in May 1862.

Darwin concentrated on orchids while battling his recurring illness. In the meantime, Huxley began to travel the country to promote Darwin's work. Nearly nine months after Darwin's orchid book was published, Huxley's own *Evidence as to Man's Place in Nature* was released. Huxley's book furthered evolutionary ideas by including his theory that man descended from apes.

As expected, the church officials took a dim view of these new theories

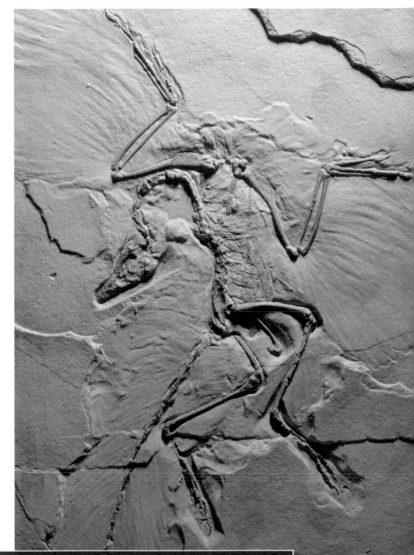

DARWIN CONSIDERED THIS FOSSIL OF *ACHAEOPTERYX*, DONATED TO THE BRITISH MUSEUM IN 1863, AS A PERFECT EXAMPLE OF TRANSMUTATION IN PROGRESS. IT IS CLEARLY A BIRD, YET IT IS IN MANY WAYS SIMILAR TO AN ANCIENT LINE OF REPTILES FROM WHICH IT IS DESCENDED.

because they challenged the foundations of their religious beliefs. So did Owen, who accidentally furthered Darwin's cause in 1863 when he gave a fossil to the British Museum. This particular fossil, called *Archaeopteryx*, appeared to be a hybrid between lizard and bird. Discovered in the Jurassic Solnhofen Limestone of southern Germany, the fossil is considered by many paleontologists to be the first bird species. Yet it has significant similarities to an ancient line of reptiles, dating back 150 million years. Paleontologists view *Archaeopteryx* as a species that was in the transitional stage between reptile and bird. In Darwin's eyes, this fossil was a perfect example of transmutation in progress.

DARWIN AS AUTHOR

Darwin was a gifted writer, able to reach both scientific and general readers. This skill certainly contributed to the positive reception of his most controversial titles, *On the Origin of Species* and *The Descent of Man*. But he also wrote many other books on a variety of topics, beginning in 1839 with the journals from his biological and geological observations and explorations on the voyage of the *Beagle*. Later books were about birds, barnacles, fossils, insects, many types of plants including carnivorous and climbing ones, human emotions, and a biography of his grandfather Erasmus Darwin. The Web site AboutDarwin.com lists twenty-five titles in all.

THE CONTROVERSY CONTINUES

Debate over the concept of evolution did not slow down. In response to the church's marshaling its forces to condemn evolution, a group of scientists founded the X Club. Its members were like-minded scientists who held discussions of current scientific theory without fear of disruption by the church or others with a non-scientific agenda. The club began publishing *Nature* magazine as an outlet to share their views with the general public. The magazine remains in print today.

Controversy continued over evolution as a frail Darwin

Contemporary cartoon by Thomas Nast.

"Gorilla: 'That Man wants to claim my Pedigree. He says he is one of my Descendants.'"

"Mr. Bergh [founder of the A. S. P. C. A.]: 'Now, Mr. Darwin, how could you insult him so?'"

FAMED CARTOONIST THOMAS NAST (CREATOR OF THE DONKEY AND ELEPHANT SYMBOLS FOR THE DEMOCRATIC AND REPUBLICAN PARTIES IN THE UNITED STATES) MEANT TO SATIRIZE DARWIN'S THEORY IN THIS AUGUST 19, 1871, CARTOON IN *HARPER'S WEEKLY* MAGAZINE. TODAY, IT MIGHT BE INTERPRETED AS FAVORING DARWIN BECAUSE MOST PEOPLE ACCEPT EVOLUTION AND THINK THE GORILLA WOULD BE EMBARRASSED BY ITS DESCENDANTS WHO REJECT DARWIN'S THEORY. (THE A.S.P.C.A. IS THE AMERICAN SOCIETY FOR THE PREVENTION OF CRUELTY TO ANIMALS.)

made a rare appearance in November 1864 to accept the Copley Medal, the highest award from the Royal Society of London. Certain factions within the Royal Society wanted to present the medal to Adam Sedgwick, Darwin's geology professor from Cambridge. But the medal was awarded to Darwin under the condition that *Origin* not be named as a factor in the decision.

Queen Victoria remained above the arguments, choosing not to get involved, and never recognizing Darwin's contributions. Several of Britain's prime ministers, including Benjamin Disraeli, did not agree with Darwin's theories. In fact, when asked where he stood on the subject of where man came from, heaven or ape, Disraeli replied that he was on the side of the angels.

Darwin remained quite ill during this period, not returning to London and the society until April 1866. A year later, he began working on a book specifically about human origins. The work grew too large for one book, so he divided it into two volumes: *The Descent of Man, and Selection in Relation to Sex* (1871). Because *The Descent of Man* described evidence that humans and the great apes had evolved from a common ancestral species, it was very controversial. He also returned to writing about other life-forms in the two-volume work *The Variation of Animals and Plants under Domestication* (1868).

Between projects and illnesses, Darwin continued to refine his work. A decade after its first publication, the fifth edition of *Origin* was released. The sixth edition, published in 1872, used the word "evolution" for the first time. With *Descent* a huge success as well, Darwin divided his time between revising the two works. In April 1874, he completed work on *Descent's* second edition. This was the last time Darwin wrote about evolution.

Darwin continued to write on a wide range of other topics, and his books remained popular until the end. His health, however, continued to deteriorate until he died on April 19, 1882. He was laid to rest in a place of great honor, in Westminster Abbey, some 20 feet (6 m) from another revolutionary English scientist: Isaac Newton.

In 1887, Darwin's son Francis published Darwin's autobiographical notes for his family along with other writings as *The Life and Letters of Charles Darwin*. However, that version left out many interesting passages that revealed Darwin's views on religious issues. Many of those deletions were restored in Darwin's granddaughter Nora Barlow's expanded version, *The Autobiography of Charles Darwin, 1809–1882*, published in 1958.

CHAPTER SIX

FROM DARWIN TO DARWINISM

*T*he theory of evolution has had impacts and led to controversies far beyond those that even Darwin and evolution's most ardent supporters could have ever imagined. In the political realm, especially in the United States but also in several other countries, supporters of creation science and intelligent design continue to attack the theory and argue that their views should be accepted and taught as science in schools. They use Thomas Henry Huxley's term "Darwinism" to characterize the science of evolution as if it were a political movement.

That term does a disservice in another way. Charles Darwin was by no means the only person responsible for developing the theory of evolution by natural selection. Alfred Russel Wallace in particular was pursuing many of the same ideas as Darwin. Without Wallace's competition, Darwin may have continued to delay the publication of *On the Origin of Species* rather than face the controversy it was certain to arouse.

Still, Darwin deserves all the acclaim that has come his way. As a naturalist, a geologist, and an author, he will be forever remembered for a book and a body of work that revolutionized the life sciences and science in general.

Since Darwin's death, controversies have continued in his name. The word "Darwinism" continues to define controversy today as it did during his lifetime. It also continues to shape modern thought in many important ways.

THE SCOPES "MONKEY TRIAL"

Even though scientists generally regard the theory of evolution as among the best supported in science, supporters of creation science and intelligent design continue to fight it in the educational system. Particularly in the United States, this has led to a series of notable trials.

The best known of those trials took place in 1925 in Dayton, Tennessee. High school science teacher John T. Scopes (1900–1970) was arrested and tried for violating the state's antievolution law that prohibited the teaching of evolution in public schools. It became known as the "Monkey Trial" because evolution opponents disputed Darwin's claims that humans and apes were close

EVEN THOUGH EVOLUTION IS ONE OF THE BEST SUPPORTED THEORIES IN ALL OF SCIENCE, TEACHING IT REMAINS THE SUBJECT OF NUMEROUS COURT CHALLENGES IN THE UNITED STATES. IN JULY 1925, TWO OF THE NATION'S LEADING ATTORNEYS, WILLIAM JENNINGS BRYAN (*LEFT, HOLDING A FAN*) AND CLARENCE DARROW (*STANDING, ARMS FOLDED*) FACED OFF IN DAYTON, TENNESSEE. AT TIMES, THE COURTROOM WAS SO HOT THAT THE TRIAL WAS MOVED OUTSIDE. TEACHER JOHN SCOPES WAS CONVICTED AND FINED FOR TEACHING EVOLUTION, THOUGH THE VERDICT WAS EVENTUALLY OVERTURNED.

relatives on the evolutionary family tree. (Humans are not as closely related to monkeys, which are a different branch of the primate family tree, but that fine distinction means little to people who dispute the whole idea of evolution.)

Nearly every newspaper in the nation and most around the world covered the story. Prosecuting the case was William Jennings Bryan (1860–1925), a three-time presidential candidate. Although Bryan had not practiced law in thirty years, he volunteered his services as prosecutor because he was a strong supporter of banning the teaching of evolution. Previously, he led the fight to ban the teaching of evolution by helping fifteen states, including Tennessee, adopt laws against it.

Famed attorney Clarence Darrow (1857–1938) headed the defense. Darrow, a defender of the "underdog" and a leader in the fight against capital punishment, had defended hundreds of people and was ready to take on Bryan.

The trial began on July 10 and quickly became a media circus. The tiny Rhea County Court House, presided over by Judge John T. Raulston, was packed with spectators, journalists, and radio broadcasters.

The most memorable event of the trial was when Darrow called Bryan to the witness stand to testify as an expert on the Bible. Darrow managed to get Bryan to concede that not everything written in the Bible should be taken literally. This acknowledgment was a huge admission given the tenor of the trial, but it didn't sway the jury. Scopes was convicted and fined $100 for teaching evolution. Darrow appealed the Scopes case before the Tennessee Supreme Court, which, in 1927, overturned the conviction on a technicality.

It was not until 1968—forty-three years after the Scopes trial—that the United States Supreme Court fully outlawed banning the teaching of evolution. In the case of *Epperson v. Arkansas*, the justices ruled that the ban was based on a particular religious teaching and thus violated the First Amendment to the U.S. Constitution.

THE BATTLES CONTINUE

Antievolution forces continue to try to influence science education in the United States, but they have lost one court battle after another. Usually the legal fights erupt over local school boards' decisions to include creation science or intelligent design as part of the science curriculum.

In 1999, the Kansas Board of Education dropped evolution from the subjects tested on state standardized tests. Kansas voters removed the antievolution members in the election of 2000 and restored the standards. In 2004, a new school board proposed to teach ID, but pro-evolution candidates won the majority in 2006.

In February 2004, Georgia's school superintendent Kathy Cox suggested removing the word "evolution" from that state's school curriculum. She suggested using the phrase "biological changes over time" to replace it. This change would teach the ideas, she argued, without the word "evolution" acting as a trigger for debate. A public outcry forced the proposal off the table.

Then it was Pennsylvania's turn. In 2005, in the most important test of the antievolution position since the Scopes trial, parent Tammy Kitzmiller, along with ten others, challenged the Dover School Board's decision to require the reading of a statement about ID in all biology classes to present an alternative to the theory of evolution. The school board also recommended that a supplementary textbook championing ID, *Of Pandas and People*, be used in biology courses to counter the views presented in the new teacher-recommended text *Biology: The Living Science*. As Brown University professor Kenneth R. Miller writes in his book *Only a Theory: Evolution and the Battle for America's Soul*, "One of the board's members complained that the [textbook *Biology: The Living Science*] was 'laced with Darwinism from beginning to end' and set about helping to present an alternative to teachers."

As one of the authors of *Biology: The Living Science*, Miller doesn't dispute Darwin's influence on its contents. He and his coauthor would have been irresponsible to approach biology in any other way, and he testified to that effect at the trial. The judge ruled that intelligent design was not science; it was religious doctrine masquerading as science. The ID book was not a scientific alternative, and the board was forced to remove it.

Someday the ID advocates may come back with better tactics. But tactics do not beat evidence. They will not be challenging an ideology called "Darwinism." Nor will they be challenging "only a theory." They will be challenging one of the pillars of modern science, Darwin's Theory of Evolution, with a capital T. And unless they can produce the intelligent designer in court, they will almost surely lose.

TIMELINE

1809 Darwin is born in Shrewsbury, England, on February 12.

1831 Darwin graduates from Christ's College, Cambridge, on April 26; leaves England aboard the H.M.S. *Beagle* on December 27.

1835 The *Beagle* reaches Galápagos Islands on September 15.

1836 Darwin returns to England after a five-year voyage on the *Beagle* on October 2.

1839 Darwin marries his cousin Emma Wedgwood. The marriage produces ten children.

1851 Darwin's beloved daughter Annie dies at age ten, shaking his Christian beliefs to their core.

1859 *On the Origin of Species by Means of Natural Selection* is published on November 22.

1860 *Origin* goes into its second printing of 3,000

1879 Darwin publishes a biography of his grandfather, titled *Life of Erasmus Darwin*, on November 19.

1881 Darwin's final book, *The Formation of Vegetable Mould Through the Actions of Worms*, is published.

1882 Darwin dies on April 19.

1925 The Scopes trial begins in Dayton, Tennessee, on July 10.

1968 In *Epperson v. Arkansas*, the U.S. Supreme Court finds that the Arkansas law prohibiting the teaching of evolution is unconstitutional.

1999 The Kansas Board of Education drops evolution from the subjects tested on state standardized tests.

2000–2006 The majority of the Kansas Board of Education changes three times from pro- to anti-evolution and back again.

2005 Parent Tammy Kitzmiller and ten other plaintiffs of Dover, Pennsylvania, are successful in a lawsuit against their school board, which had required that a statement be read about intelligent design in all biology classes and placed an ID textbook in the library as an alternative to standard biology textbooks. The judge ruled that ID was religious doctrine, and thus the book was removed as a violation of the constitutional prohibition on state promotion of a particular religion.

2013 More than 1,200 personal letters that Darwin wrote to his longtime friend Joseph Hooker are published by Cambridge University's Darwin Correspondence Project and are made available online.

GLOSSARY

BOTANY The branch of biology that studies plant life.

CORAL A polyp-like sea animal that lives in colonies of vast numbers of individuals. After they die, their outer skeletons accumulate to form reefs.

CREATIONISM The belief in the literal interpretation of the biblical account of the creation of the universe.

CROSSBREEDING The mating of individual animals of different breeds, varieties, or species to produce hybrid offspring.

ENTOMOLOGIST A person who studies insects.

ENVIRONMENT The habitat that particular species live in and adapt to.

FINCH A songbird typically having a short, stout bill. Darwin studied differences in finch species from different Galápagos islands to understand how species transmutate, or evolve.

FOSSIL A remnant of an organism, such as a skeleton or leaf imprint, embedded and preserved in a rock.

GEOLOGY The scientific study of the origin, history, and structure of the earth.

HYBRID A mixed species formed by the breeding of two or more different species.

HYBRIDIZATION The process of interbreeding two different species.

HYPOTHESIS An educated guess to explain observed facts, evidence, or relationships in a particular field of study. If further evidence supports the hypothesis, it may become the basis of a theory.

NATURALIST A person who studies the life and geology of the natural world.

NATURAL SELECTION The process by which a species changes through reproducing traits that are beneficial in its environment.

PALEONTOLOGIST A scientist who studies fossil plants and animals.

QUARANTINE A waiting period necessary to make sure that a person has not been exposed to a contagious disease.

SPECIES An organism belonging to a group with similar biological traits.

TAXIDERMY The practice of stuffing and mounting the skins of animals.

THEORY A well-supported idea or set of ideas to explain observed facts, evidence, or relationships in a particular field of study.

TRANSMUTATION The process of changing in form. In biology it describes how one species transforms into another.

ZOOLOGY The branch of biology that studies animals and animal life, including the study of the structure, physiology, development, and classification of animals.

FOR MORE INFORMATION

American Museum of Natural History
Central Park West at 79th Street
New York, NY 10024-5192
(212) 769-5100
Web site: http://www.amnh.org

The American Museum of Natural History is a world-
renowned scientific and cultural institution and
was founded in 1869. The museum held a special
exhibition on Charles Darwin, his voyage on the
Beagle, and his studies on evolution in 2005–2006
(see http://www.amnh.org/exhibitions/past-exhi-
bitions/darwin).

California Museum of Paleontology
University of California
1101 Valley Life Sciences Building
Berkeley, CA 94720-4780
(510) 642-1821
Web site: http://www.ucmp.berkeley.edu

The museum's mission is "to investigate and promote
the understanding of the history of life and the
diversity of the Earth's biota through research and
education."

Charles Darwin Foundation for the Galápagos Islands
Puerto Ayora, Santa Cruz Island
Galápagos, Ecuador
(593) 5 2526-146/147

Web site: http://www.darwinfoundation.org
The foundation "provides scientific research and tech-
  nical information and assistance to ensure the
  proper preservation of the Galápagos Islands. "

Down House
Luxted Road
Downe, Kent BR6 7JT
England
0 1689 859119
Web site: http://www.english-heritage.org.uk/daysout
  /properties/home-of-charles-darwin-down-house
Down House is the home of Charles and Emma Darwin.
  In the study of this house, Darwin wrote *On the
  Origin of Species*.

Natural History Museum
Cromwell Road
London SW7 5BD UK
England
+44 (0)20 7942 5000

Web site: http://www.nhm.ac.uk
This museum offers information about the natural
  world through its collections and strives to educate
  people about preserving our planet. For additional
  information, check out the Darwin Center Web site
  (http://www.nhm.ac.uk/visit-us/darwin-centre-
  visitors/index.html) and the Charles Darwin and

HMS *Beagle* Web site (http://www.nhm.ac.uk/
research-curation/science-facilities/library/themes/
travel-and-exploration/hms-beagle/index.html).

National Museum of Natural History
Smithsonian Institution
Constitution Avenue NW
Washington DC 20560
(202) 633-1000
Web site: http://www.mnh.si.edu
The National Museum of Natural History (NMNH) is part
of the Smithsonian Institution and exhibits artifacts
and collections on natural history and cultures from
around the world. It also provides research materials and offers educational programs to the public.
The NMNH's Web page (http://humanorigins.si.edu)
contains interesting information about evolution
and fun facts about the origins of humankind.

WEB SITES

Due to the changing nature of Internet links, Rosen
Publishing has developed an online list of Web sites
related to the subject of this book. This site is updated
regularly. Please use this link to access the list:

http://www.rosenlinks.com/RDSP/darw

FOR FURTHER READING

Ashby, Ruth. *Young Charles Darwin and the Voyage of the* Beagle. Atlanta, GA: Peachtree, 2009.

Burkhardt, Frederick. *Origins: Selected Letters of Charles Darwin 1822–1859*. New York, NY: Cambridge University Press, 2008.

Eldredge, Niles, and Susan Pearson. *Charles Darwin and the Mystery of Mysteries*. New York, NY: Roaring Brook Press, 2010.

Johnson, Sylvia A. *Shaking the Foundation: Charles Darwin and the Theory of Evolution*. Minneapolis, MN: Twenty-First Century Books, 2013.

Krull, Kathleen, and Boris Kuliyov. *Charles Darwin*. New York, NY: Viking, 2010.

Leone, Bruno. O*rigin: The Story of Charles Darwin*. Greensboro, NC: Morgan Reynolds Publishing, 2009.

Lew, Kristi. *Evolution: The Adaptation and Survival of Species*. New York, NY: Rosen Publishing, 2011.

Nardo, Don. *The Theory of Evolution: A History of Life on Earth*. Minneapolis, MN: Compass Point Books, 2010.

Schanzer, Rosalyn. *What Darwin Saw: The Journey That Changed the World*. Washington, DC: National Geographic, 2009.

Winston, Robert M. L. *Evolution Revolution*. New York, NY: DK Publishing, 2009.

BIBLIOGRAPHY

AboutDarwin.com. "Dedicated to the Life and Times of Charles Darwin." Retrieved March 22, 2013 (http://www.aboutdarwin.com).

All About Science. "Darwin's Theory of Evolution." Retrieved January 2004 (http://www.darwins-theory-of-evolution.com).

Bowler, Peter J. *Evolution: The History of an Idea.* Berkeley, CA: University of California Press, 2003.

Cornell University Law School, Legal Information Institute. *Epperson v. Arkansas* (No. 7). Retrieved March 22, 2013 (http://www.law.cornell.edu/supct/html/historics/USSC_CR_0393_0097_ZS.html).

Coyne, Jerry A. *Why Evolution Is True.* New York, NY: Viking, 2009.

Darwin, Charles. *The Autobiography of Charles Darwin 1809–1882.* New York, NY: W.W. Norton & Co., 1993.

Darwin, Francis, ed. *The Life and Letters of Charles Darwin.* Retrieved March 22, 2013 (http://www.victorianweb.org/science/darwin/darwin_autobiography.html).

Dawkins, Richard. *The Ancestor's Tale: A Pilgrimage to the Dawn of Evolution.* Boston, MA: Houghton Mifflin, 2004.

Dawkins, Richard. *River Out of Eden: A Darwinian View of Life.* New York, NY: Basic Books, 1995.

Eldredge, Niles. *Reinventing Darwin: The Great Debate at the High Table of Evolutionary Theory*. New York, NY: John Wiley & Sons, 1995.

Gribbin, John, and Mary Gribbin. *FitzRoy: The Remarkable Story of Darwin's Captain and the Invention of the Weather Forecast*. New Haven, CT: Yale University Press, 2004.

Indiana University of Pennsylvania. "Darwin's Theory of Evolution." Retrieved January 2004 (http://nsm1.nsm.iup.edu/rwinstea/darwin.shtm).

Kevles, Daniel. "In the Name of Darwin." Retrieved March 22, 2013 (http://www.pbs.org/wgbh/evolution/darwin/nameof/index.html).

Keynes, Richard Darwin. *Fossils, Finches, and Fuegians: Charles Darwin's Adventures and Discoveries on the Beagle, 1832–1836*. London, England: HarperCollins, 2002.

LucidCafé. "Charles Darwin: British Naturalist." Retrieved January 2004 (http://www2.lucidcafe.com/lucidcafe/library/96feb/darwin.html).

Miller, Kenneth R. *Only a Theory: Evolution and the Battle for America's Soul*. New York, NY: Viking Penguin, 2008.

Online Literature Library. "Charles Darwin." Retrieved January 2004 (http://www.literature.org/authors/darwin-charles).

Sobel, Dava. *Longitude: The True Story of a Lone Genius Who Solved the Greatest Scientific Problem of His Time.* New York, NY: Walker, 1996.

SpaceandMotion.com. "Charles Darwin: The Theory of Evolution." Retrieved January 2004 (http://www.spaceandmotion.com/Charles-Darwin-Theory-Evolution.htm).

Stott, Rebecca. *Darwin and the Barnacle: One Tiny Creature and History's Most Spectacular Scientific Breakthrough.* New York, NY: W. W. Norton & Co., 2003.

TalkOrigins.org. "Evolution FAQs." Retrieved January 2004 (http://www.talkorigins.org/origins/faqs-evolution.html).

University of California, Berkeley, Museum of Paleontology. "Evolution Entrance." Retrieved March 2013 (http://www.ucmp.berkeley.edu/history/evolution.html).

Vermeij, Geerat. *The Evolutionary World: How Adaptation Explains Everything from Seashells to Civilization.* New York, NY: Thomas Dunne Books, 2010.

INDEX

ABOUT THE AUTHOR

After his third book, *Catastrophe! Great Engineering Failure—and Success*, was designated a "Selector's Choice" on the 1996 list of Outstanding Science Trade Books for Children, physicist Fred Bortz decided to become a full-time writer. His books, now numbering nearly thirty, have since won awards, including the American Institute of Physics Science Writing Award, and recognition on several best books lists.

Known on the Internet as the smiling, bowtie-wearing "Dr. Fred," he welcomes inquisitive visitors to his Web site at www .fredbortz.com.

PHOTO CREDITS